THE ELEMENTS

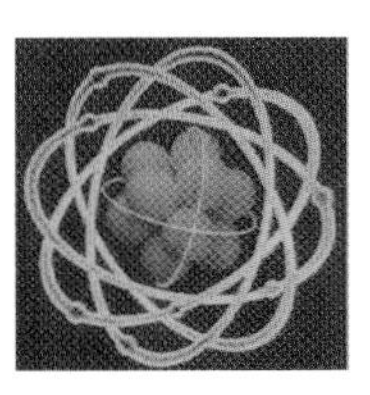

Silicon

Jens Thomas

BENCHMARK BOOKS
MARSHALL CAVENDISH
NEW YORK

Benchmark Books
Marshall Cavendish Corporation
99 White Plains Road
Tarrytown, New York 10591

Library of Congress Cataloging-in-Publication Data
Thomas, Jens.
Silicon / Jens Thomas.
p. cm. — (The elements)
Includes index.
ISBN 0-7614-1274-3
1. Silicon—Juvenile literature. [1. Silicon.] I. Title. II. Elements (Benchmark Books)
QD181.S6 T52 2001
546'.683—dc21 2001025993

Printed in Hong Kong

Picture credits

Front cover: Reinhard Janke/Still Pictures.
Back cover: Pictor International.
Corbis: José Manuel Sanchis Calvete 15; Leif Skoogfors 30.
Ecoscene: Richard Glover 10 (below).
Hulton Archive: 14.
Image Bank: Lisl Dennis i, 11; Will Crocker iii, 23.
Pictor International: 4, 10 (above), 22, 24, 25, 27.
Science Photo Library: Andrew McClenaghan 6;
Clive Freeman, The Royal Institution 19 (below); Northwestern University 7;
Roberto de Gugliemo 19 (above); Volker Steger, Peter Arnold Inc. 26.
Still Pictures: Reinhard Janke 5.
Travel Ink: David Toase 9.
TRIP: D. Maybury 20.
Wacker Siltronic: 12, 16.

Series created by Brown Partworks Ltd.
Designed by Sarah Williams

Contents

This picture of sand has been magnified, so you can see the individual sand grains. Each one is a tiny piece of quartz, which contains many silicon atoms.

What is silicon?

You might not realize it, but you are surrounded by silicon all the time. You tread on this element whenever you walk on the ground—it can be found in almost all the rocks, clays, and sands that form Earth's crust. Silicon also makes up a large part of man-made structures such as bricks and glass. In addition, it is extremely important to the electronics industry. The electronic chips that make modern-day computers, calculators, and even refrigerators work would not exist without this element. Silicon also occurs inside your body, where it may help to strengthen your bones and skin.

Although silicon is found everywhere on Earth, it always occurs together with other elements in compounds. It is these silicon compounds that are so useful in our lives.

The silicon atom

Silicon has the chemical symbol Si. Like all elements, it is made up of tiny particles called atoms. The atoms are made up of much smaller particles called protons and neutrons, which form a core—or nucleus—at the center of the atom. Even tinier particles, called electrons, revolve around the nucleus.

Protons have positive charge and electrons have negative charge. Silicon has

SILICON ATOM

Nucleus
First shell
Second shell
Third shell

Every silicon atom contains 14 electrons. These negatively charged particles orbit the nucleus in three shells. The first shell contains two electrons, the second shell contains eight electrons, and the outermost shell contains four electrons.

DID YOU KNOW?

SILICON ISOTOPES

The number of protons and neutrons in an atom gives the element its atomic mass. Silicon isotopes have different atomic masses, but they still behave the same chemically, because they have the same numbers of protons and electrons. More than 92 percent of silicon atoms on Earth have 14 neutrons in their nucleus, and these have an atomic mass of 28. However, some have 15 neutrons—and there are even a few with 16. As a result, silicon has an average atomic mass of 28.1.

an atomic number of 14, which tells us that each atom of silicon has 14 protons in its nucleus. Most of the time, atoms in an element have equal numbers of protons and electrons, so each silicon atom has 14 electrons. Neutrons get their name because they are neutral (they have no charge).

This means that atoms of the same element can have different numbers of neutrons without affecting the atoms' charge. Atoms that have the same numbers of protons and electrons but different numbers of neutrons are called isotopes.

This picture shows the manufacture of a tiny "chip" made from a wafer of pure silicon. Silicon chips are used in a huge array of modern electronic devices.

Special characteristics

Pure silicon is a solid that can exist in two forms. One form of silicon is made up of shiny black crystals, and the other form is a dull brown powder. Both types have very high melting and boiling points—they melt at 2,570°F (1,410°C) and boil at 4,270°F (2,355°C). Silicon is a heavy element and has a density of 2.33 grams per cubic centimeter. This means that a lump of silicon weighs 2.33 times more than an identical volume of water.

Silicon has a hardness of 6.5 on the Mohs scale—it is harder than glass but softer than a steel file. The hardest material is diamond, with a hardness of 10.

Sharing electrons

One of the reasons that silicon is found in so many places and is so important is that it can undergo reactions and form a large number of different chemical compounds. It can do this because of the number of electrons it contains in its atoms.

Silicon lies in Group IV of the periodic table. Like other elements in this group, silicon needs eight electrons to fill its

DID YOU KNOW?

SILICON AS A METALLOID

Almost all the elements in the periodic table are either metals or nonmetals. The metals are shiny, and most of them are hard but can be bent or hammered into different shapes without breaking. These elements are also very good at conducting electricity. Most of the elements in the periodic table are metals. By contrast, nonmetals break if they are stretched or hammered, and they do not conduct electricity. Silicon is one of the very few elements that does not fit easily into one of these two categories. It is hard and breakable like a nonmetal but can conduct electricity a little like a metal. For this reason, silicon is described as a semimetal, or a metalloid.

outermost shell and make its atoms stable. It has only four electrons in its outer shell, however, so it is four "short" of the number it needs for stability. Silicon can get these "missing" electrons by sharing electrons with other atoms. Sharing electrons in this way creates a chemical bond between atoms, allowing each of them to fill their outer electron shell. Silicon can form bonds with four different atoms at the same time, so it can form many more compounds than can most other elements.

Silicon forms large crystals in which the atoms line up in ordered patterns. This image is blurred because it has been photographed at very high magnification.

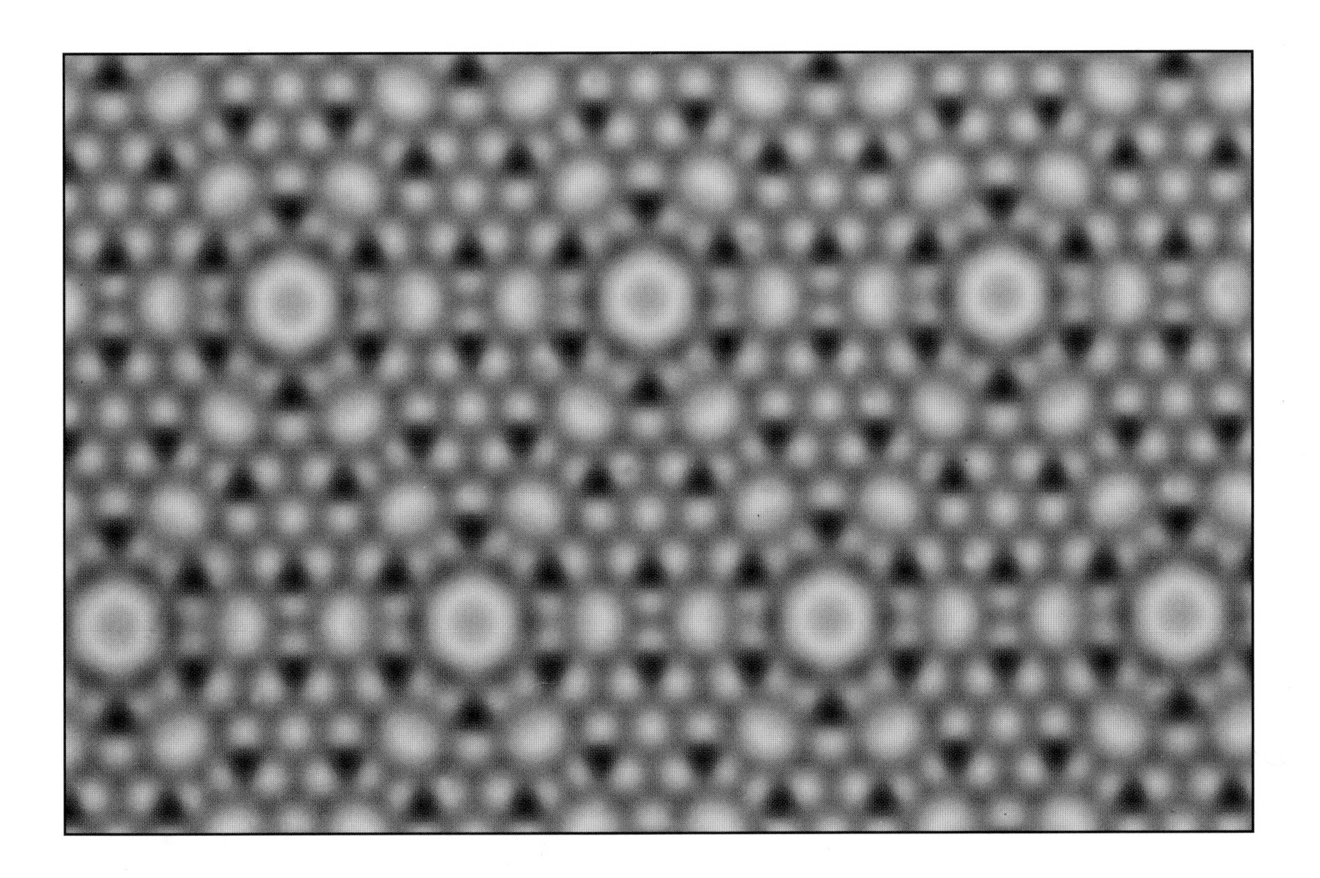

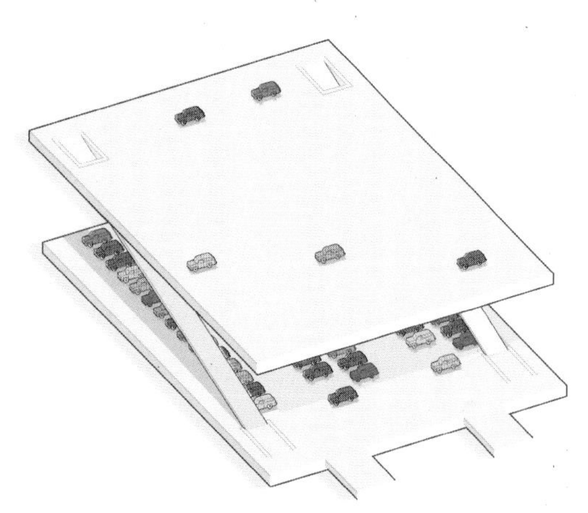

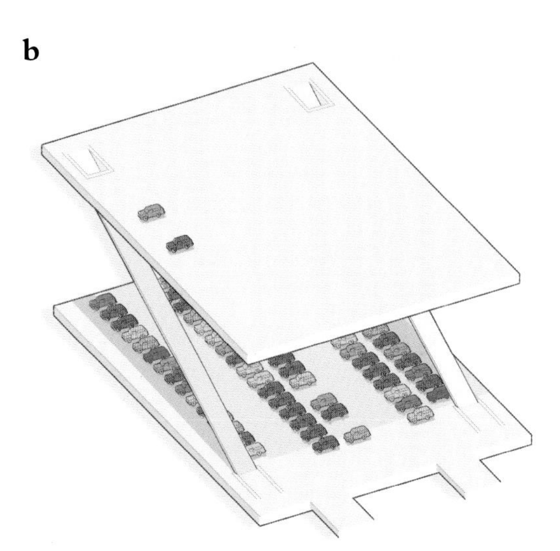

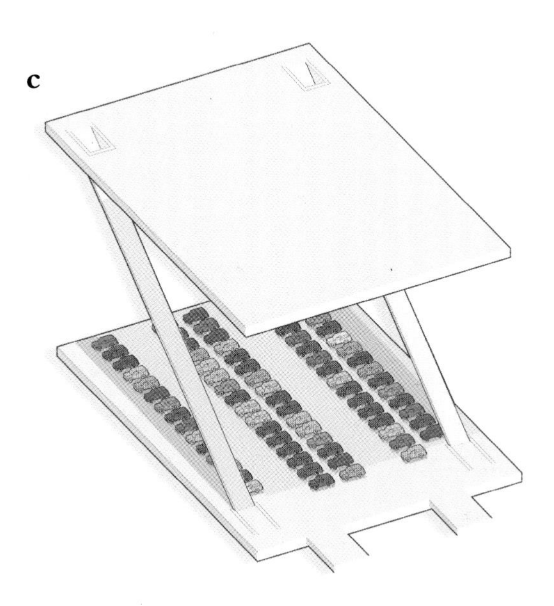

Imagine the conduction band as the upper level in a multi-level parking garage. In conductors (a) the levels are close together, so many electrons (cars) can move up to the upper level. In semiconductors (b) a few electrons can get to the upper level, but in insulators (c) the gap is too big.

Conducting electricity

Metals are good electrical conductors because the electrons that carry the electricity are not tightly tied to their atoms. Instead, they can roam through the metal in a kind of "electricity highway" called the conduction band. In nonmetals, the electrons are held very tightly to the atoms and are not able to enter the conduction band. This prevents nonmetals from conducting electricity, and these elements are described as insulators.

At low temperatures, silicon is an insulator. The reason for this is that the four electrons in the outer shell of an atom of solid silicon are held tightly in bonds with neighboring silicon atoms. However, if the electrons in silicon are given extra energy—in the form of heat, for example—some of them can break free from the bonds and enter the conduction band. Then they can move through the solid silicon, carrying electricity. For this reason, silicon is called a semiconductor—under the right conditions it does conduct electricity, but not as well as a metal does.

Where silicon is found

Silicon occurs almost everywhere. It is the seventh most abundant element in the Universe and can be found in many stars, including the Sun. Here on Earth, silicon is the second most abundant element. For every four atoms that make up Earth's outer crust, more than one of them is a silicon atom.

DID YOU KNOW?

SILICON IN YOU

Silicon is very useful to living organisms, because of the chemical bonds that this element can form. Most silicon inside humans and other animals lies in the bones, the arteries, the tendons, and the skin. Scientists think that it might play an important part in strengthening these tissues. If all the silicon in your body was piled together and weighed, you would probably find that you have around ⅙ to ⅓ of an ounce (5 to 10 grams) of it.

Silica and silicates

In Earth's crust, silicon is almost always found joined to the most abundant element on the planet—oxygen. Together, silicon and oxygen form silica, which can exist in numerous forms. Quartz is probably the best-known type of silica. When silica is combined with at least one other element, a silicate mineral is formed. Minerals are naturally occurring solids in which the atoms are repeated in the same arrangement throughout the whole structure.

The name silicon comes from the Latin words *silex* or *silicis*, which mean "flint" or "rock." This is because most rocks are made up of several kinds of silicates. There are more than 1,000 different silicates, and together they make up 95 percent of Earth's crust. Common silicate minerals include olivine, asbestos, and clay.

You can see clearly the silicate minerals that make up this piece of granite. Feldspar gives the rock an overall pink color, speckled with the pale gray of quartz and the blackish browns and greens of mica and amphibole.

The gemstone amethyst is a type of quartz. It gets its beautiful and distinctive lilac and purple colors from the traces of manganese and iron it contains.

DID YOU KNOW?

BEAUTIFUL GEMSTONES

Even though most silicate minerals are common and exist abundantly in the earth, a few are valued as gemstones. Amethyst is a beautiful purple gemstone, made up of a form of quartz. Other silicate gemstones are emerald, which gets its deep green color from the chromium and iron atoms it contains, and golden-brown topaz, which contains aluminum and iron.

Silicate structure

Clear quartz crystal is the purest form of silica, but there are other forms, including sand, sandstone, and flint. The various types of silica look different from each other because they have different structures—some consist of one giant

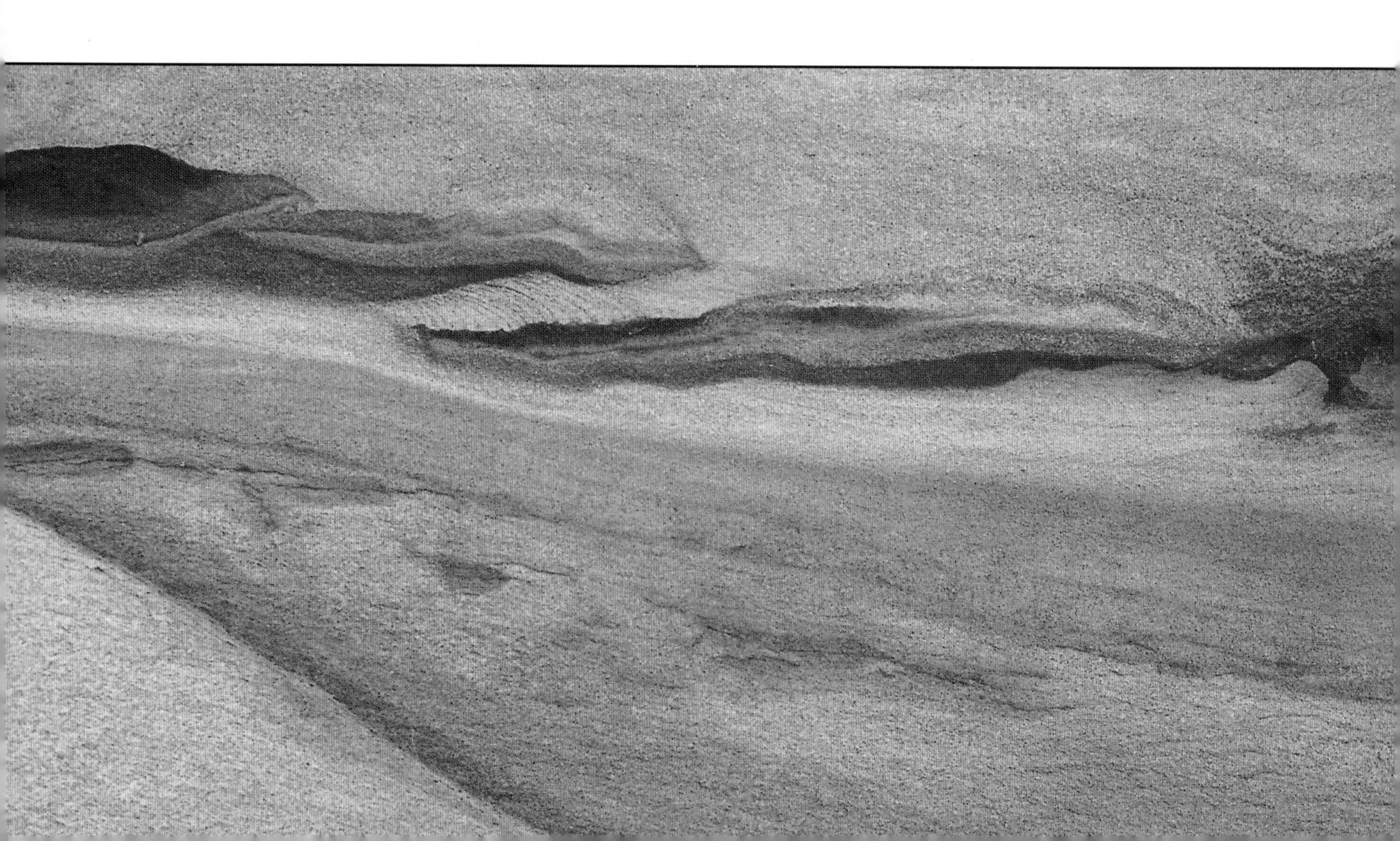

The striped appearance of sandstone shows how it has been built up from layers and layers of sediment in Earth's crust. This rock is 81 percent silica.

crystal, while others are made up of many tiny crystals. In some forms, traces of other liquids and minerals are trapped inside microscopic holes in the silicas, giving the silicas their characteristic color. Most sand is simply small pieces of silica that have been ground down by the weather and by waves on the shore.

Most silicate minerals are not only made of silicon and oxygen, however—they also contain other atoms. Aluminum, sodium, magnesium, and calcium atoms are common in silicates.

Many silicates are very useful to people. For example, quartz sand is used in making glass and house bricks, while clays are used in ceramics (pottery).

These ceramic pots are made from micaceous clay (clay that contains mica, a type of silicate mineral).

SEE FOR YOURSELF

MAKING CLAY

Clay is a silicate mineral made up of many tiny, flat particles. When water is added to these particles, they stick together but are able to slide over each other. As a result, wet clay is soft and slippery. You can make a type of clay by mixing together 1½ cups of baking soda, 2 to 3 cups of cornstarch, and 2 cups of warm water. Ask an adult to help you stir the mixture in a pan over low heat until it thickens. Although your "clay" is not made of silicate, it is made of many similar tiny particles. When the mixture cools, you can mold it into shapes, and it will harden as it dries—just like real clay does.

Extracting silicon

Pure silicon does not occur naturally on Earth, so it needs to be separated from its compounds before it can be used. Almost all silicon used today comes from a type of quartz sand, which is heated with carbon in a furnace to over 5,400°F (3,000°C). At this temperature, the carbon removes the oxygen from the silica to form the gas carbon monoxide (CO). This leaves liquid silicon at the bottom of the furnace. The silicon made this way is about 98 to 99 percent pure.

Purifying silicon

The silicon used for electronics needs to be purified further. First, it is turned into trichlorosilane ($HSiCl_3$) by mixing it with hydrochloric acid (HCl). Trichlorosilane still contains impurities, but these can be removed in a process called distillation. In this process, liquid trichlorosilane is heated to its boiling point, at which stage it changes into a gas. The impurities boil at a different temperature from that of the trichlorosilane. It is possible to make sure that only trichlorosilane gas boils off by carefully controlling the temperature. The silicon atoms can then be recovered from the pure gas.

Zone refining

The silicon produced by distillation is not pure enough for some purposes, and more purification is needed. This is done by feeding the end of a silicon rod into an electrical heater so that a small section of

Silicon is refined on a huge scale so that industries all over the world can use pure silicon in the manufacture of computers and other electronic goods.

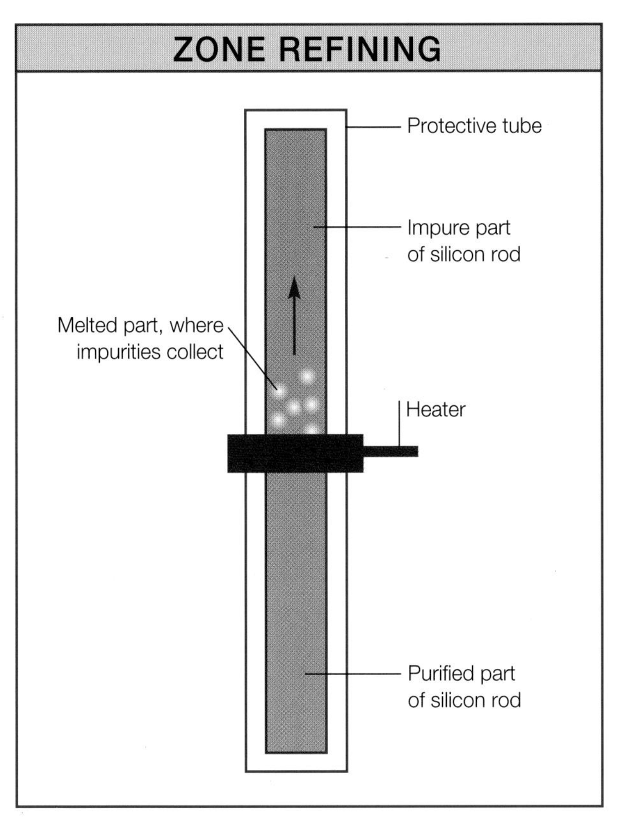

Zone refining—in which impurities are concentrated at one end of a rod and removed—is not restricted to silicon purification. This method is used to purifiy a wide range of solids, including metals.

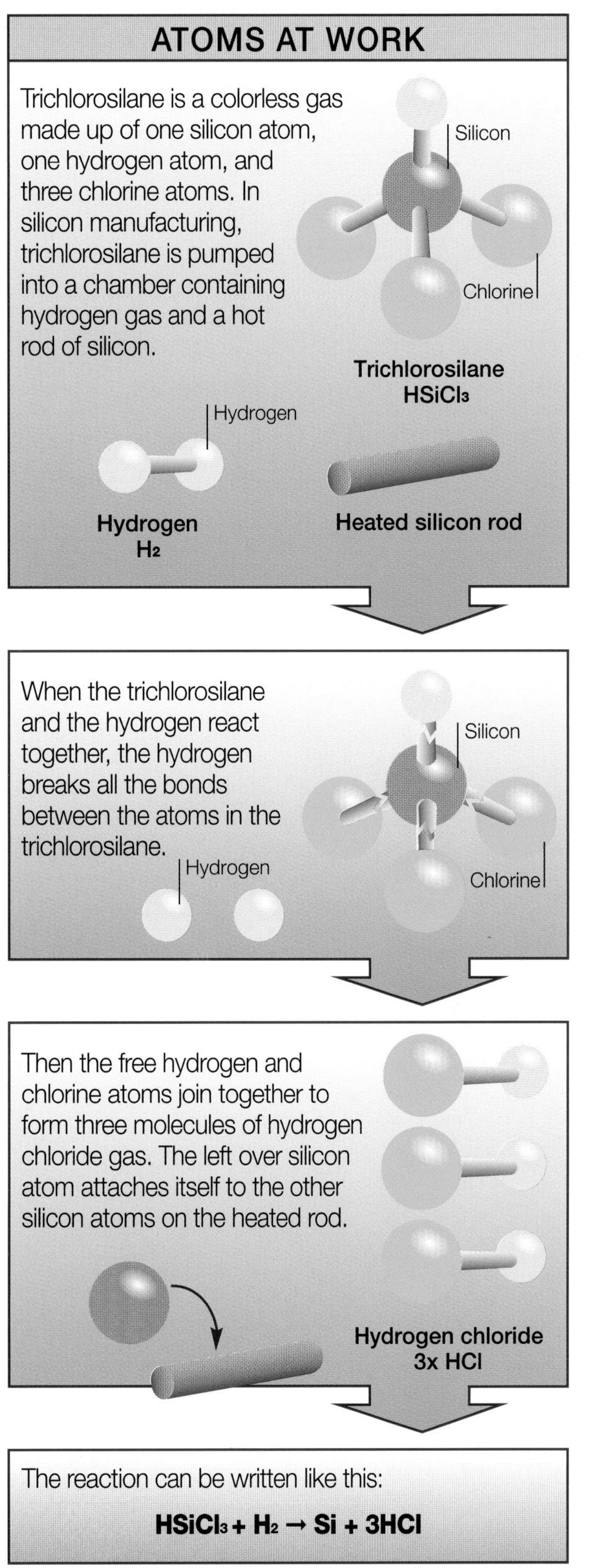

the rod melts. The rod is then moved through the heater so that a "zone" of molten silicon travels along the rod. Because the impurities in the silicon dissolve in the hot liquid, they stay in the molten zone as it travels down the rod. When the zone gets to the other end, almost all the impurities have collected there, and it can be cut off and thrown away. Silicon that has been purified in the process of zone refining can be up to 99.999999999 percent pure.

How silicon was discovered

The first silicon was probably made by French chemists Joseph-Louis Gay-Lussac (1778–1850) and Louis-Jacques Thenard (1777-1857) in around 1811. They heated the metal potassium with a colorless gas called silicon tetrafluoride (SiF_4). The heat caused the metal and the gas to react together to form silicon.

Berzelius (shown above) is known as one of the founding fathers of modern chemistry. He analyzed many compounds in detail and measured the atomic weights of 43 elements.

> **DID YOU KNOW?**
>
> ***AN IMPORTANT CHEMIST***
> Berzelius is a very important figure in the history of chemistry. He suggested the modern symbols and formulas that are still used to describe chemical reactions today. During his career, Berzelius also discovered the elements cerium (in 1804), selenium (in 1817), and thorium (in 1828).

However, Gay-Lussac and Thenard did not realize that this is what they had made.

It was left to Swedish chemist Jöns Jacob Berzelius (1779–1848) to make pure silicon and to receive the credit for discovering a new chemical element. In 1823, Berzelius made brown powdery silicon in the same way that Gay-Lussac and Thenard had done. Berzelius, however, went a stage further. He purified his silicon by carefully washing away the other chemicals left over from the reaction.

Reacting with sodium

The shiny, black silicon crystals were not prepared for the first time until 1854. This discovery was made by another French chemist—Henri Étienne Sainte-Claire Deville (1818–1881). He made the crystals by reacting a silicon compound with the metal sodium. This was the same method that Deville had already used to prepare some of the first pure samples of aluminum, boron, platinum, and titanium.

How silicon reacts

Although they are both made only of silicon atoms, the two forms of silicon look very different because of the way their atoms are connected. Within the shiny black crystals, the atoms are joined together in the same regular way throughout the whole crystal. In the powdery, dull brown silicon, the atoms are joined together in small groups. Instead of forming one huge crystal, they form many tiny microcrystals. Together, these microcrystals appear as a powder.

Despite the fact that pure silicon is never found on Earth, it very rarely reacts with other elements. Silicon is so unreactive that even the strongest acids usually have no effect on it. One exception to this rule is a very powerful acid called hydrofluoric acid, which reacts with silicon to form hexafluorosilicic acid (H_2SiF_6).

Forming compounds

There are a few situations in which pure silicon will react, however. For example, it reacts with alkalis, which is a group of chemicals that includes sodium hydroxide (bleach). Alkalis and silicon react to form silica (SiO_2). Powdered silicon also reacts with elements in a group called the

One of the reactions that silicon commonly undergoes is with alkalis to form silica (SiO_2). Given enough time for very slow crystallization, and space to grow in Earth's crust, the silica takes the form of large, clear crystals of quartz (shown below).

halogens, which contains chlorine and fluorine. When silicon and the halogens react together, they form compounds called tetrahalides.

Silicon can form many other compounds, but these compounds are usually made using silica or a tetrahalide as a starting point. Hydrides of silicon (silicon bonded to hydrogen) can be made in several ways. For example, magnesium silicide (Mg_2Si) reacts with dilute hydrochloric acid (HCl) to form many different silane compounds. The simplest is silane (SiH_4), which is a silicon atom bonded to four hydrogen atoms. The other

ATOMS AT WORK

One of the few reactions that pure silicon undergoes is with halogen elements such as chlorine. Chlorine gas molecules, each made up of two chlorine atoms, are passed over glowing red-hot silicon.

Silicon

Chlorine

Silicon
Si

Chlorine
2x Cl_2

Two molecules of chlorine gas break apart to form four chlorine atoms. These can then attach themselves to the glowing silicon, forming silicon tetrachloride. Silicon tetrachloride is a gas that is used by the military to create smoke screens.

Chlorine

Silicon

Silicon tetrachloride
$SiCl_4$

The reaction can be written like this:

$$Si + 2Cl_2 \rightarrow SiCl_4$$

Industrial plants like this one produce thousands of tons of silane (SiH_4). This gas is used in the manufacture of glass coatings and parts for photocopying machines, as well as of silicon chips.

silanes formed in this reaction consist of chains of silicon atoms attached to varying numbers of hydrogen atoms. Some of these silane chains contain as many as six silicon atoms and fourteen hydrogen atoms.

Reacting with oxygen

The most important element with which silicon forms compounds is oxygen. One of the products of a reaction between water and a silicon tetrahalide or silane is silica. The chemical formula for silica (SiO_2) might make you think that the silicon shares two electrons with each oxygen atom, but actually each silicon atom is bonded to four oxygen atoms. This means that each oxygen atom has a "spare"

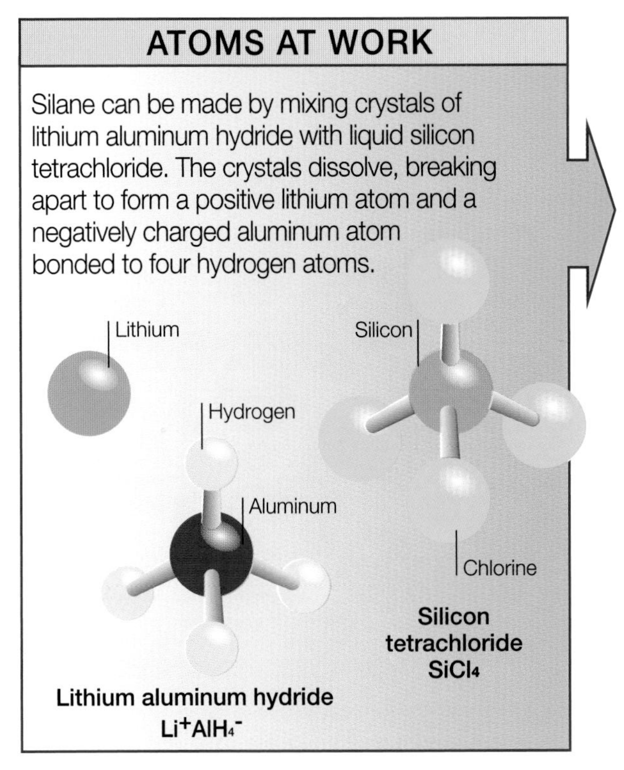

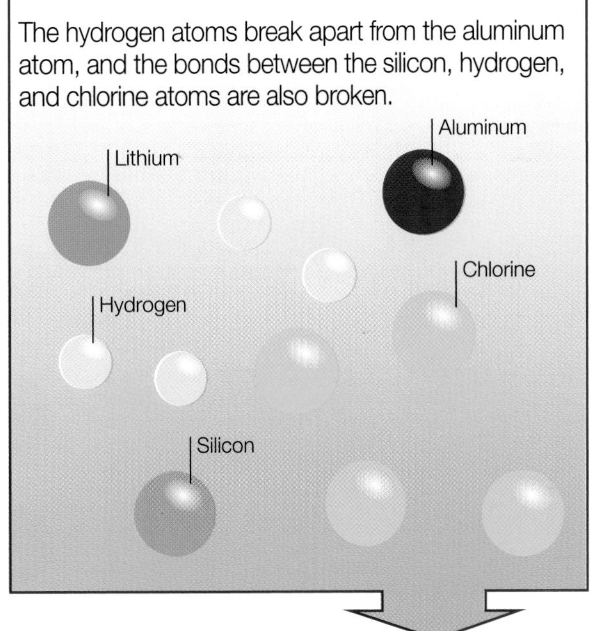

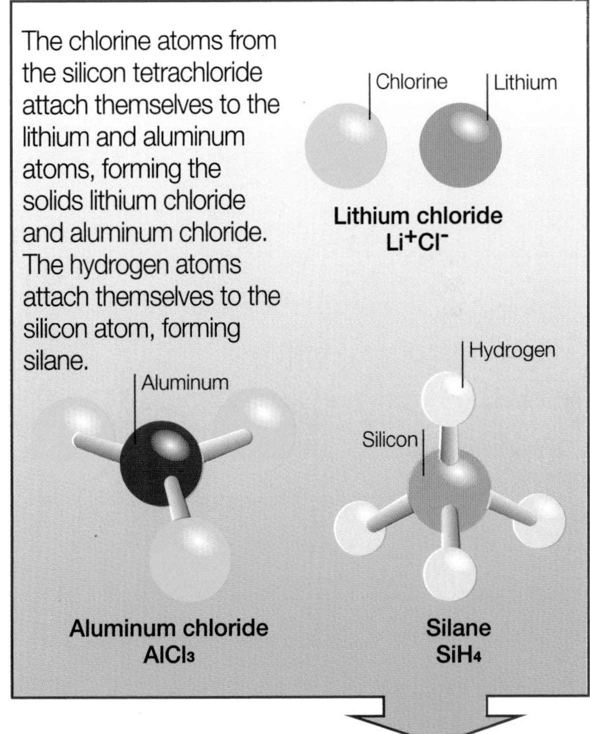

The reaction can be written like this:

$$SiCl_4 + Li^+AlH_4^- \rightarrow SiH_4 + Li^+Cl^- + AlCl_3$$

OLIVINE

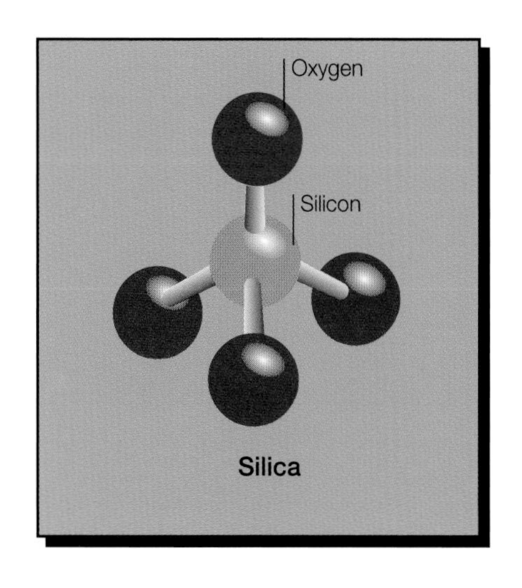

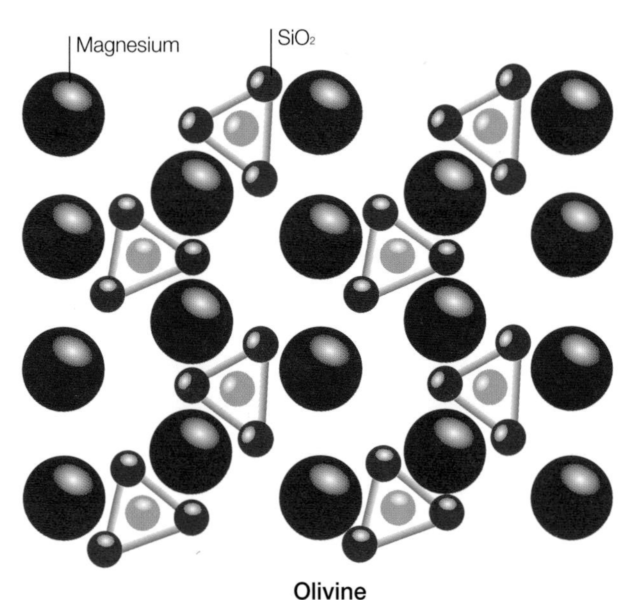

One of the thousands of silicates that can be created when silica and atoms of other elements join up together is olivine. This is a hard, glassy mineral that makes up most of Earth's mantle (the part of the planet between the core and crust). It is one of the simplest silicates and is formed when rows of magnesium or iron atoms are joined to the silica "building blocks."

electron, which can be used to bond to another silicon atom—or even an atom from a completely different element. The silicates are made up of millions of these "building blocks" connected to each other. The blocks can be connected together in different ways, and atoms of other elements can also be fitted into the spaces between them. As a result, thousands of different types of silicates can exist.

Types of silicate

When the silica blocks are joined together in a giant framework, the crystal quartz is formed. If separate blocks are packed together with atoms of iron and magnesium, the mineral olivine is created. Olivine is a common mineral on Earth and can also be found in many meteorites (rocks that fall to Earth from space). The silicate mineral beryl forms when rings made of six of the blocks are packed together with beryllium and aluminum atoms. Popular forms of beryl are the gemstones emerald and aquamarine.

Zeolites are an important form of silicate. They are made up of silica building blocks together with aluminum atoms. They also contain many other elements—and usually water, too. For this

DID YOU KNOW?

SIEVES AND SOFTENERS

The zeolites are very large minerals, and many of them contain holes and channels that are big enough to let certain molecules through but not others. As a result, these zeolites can be used as molecular "sieves," filtering out unwanted chemicals. Zeolites are also used to soften water.

Water is described as "hard" when it contains too much calcium. This can be a problem, because the calcium prevents soap from lathering properly, and it also clogs up pipes and teakettle elements (the part inside the teakettle that heats up the water). Zeolites with sodium atoms within their cavities can be used to soften water. The hard water is run through the zeolites, and the calcium ions in the water replace the sodium atoms in the zeolites, softening the water in the process.

This emerald was mined in Pindobacu, Brazil. It gets its bright green color from chromium atoms; other forms of beryl are pink and pale bluish green.

reason, many of them have extremely complicated chemical formulas. For example, a zeolite called chabazite has the formula $Ca_6Al_{12}Si_{24}O_{72}.40H_2O$, and faujasite has the amazing formula $Na_{13}Ca_{11}Mg_9K_2Al_{55}Si_{130}O_{384}.235H_2O$.

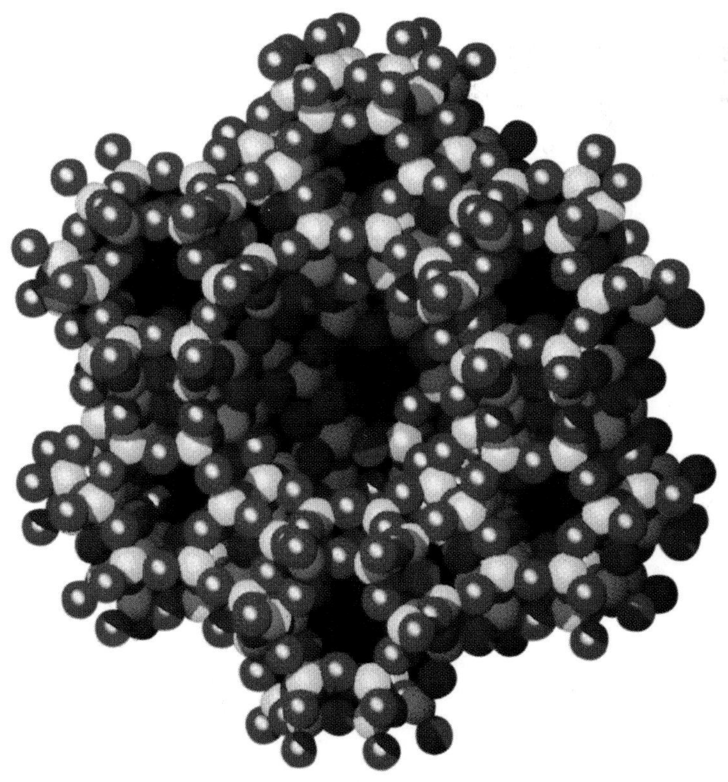

In this computer graphic of the zeolite faujasite, you can see the holes that allow it to be used as a "molecular sieve." Oxygen atoms are shown in red, while silicon and aluminum atoms are in yellow.

Silicon and carbon

All life on Earth is based on the element carbon. Carbon is extremely useful for making living things because this one element can form millions of different compounds. The carbon atoms can also link together to form long chains. The chains are important because they can join up to make cells, which contain the chemicals essential for living processes.

Plants and animals look different, but their chemistry is based on the same element—carbon. Carbon and silicon have many characteristics in common.

Similar elements

Silicon is the element just below carbon in Group IV of the periodic table. Elements that are above or below each other in the periodic table are often very similar, and this is true for silicon and carbon. Like carbon atoms, silicon atoms can each form four bonds, and they are also able to link up with each other to form long chains—although silicon has a greater tendency to form chains in which silicon and oxygen atoms alternate. In addition, silicon forms bonds with hydrogen, the halogens, and carbon. These are all elements that exist on Earth and need to be used by any form of life. The similarities between silicon and carbon have made some scientists wonder whether life based on silicon atoms instead of carbon atoms would be possible.

There is one very important reason why no life on Earth is based on silicon, however. Although silicon can form chains and compounds with hydrogen, it forms bonds much more easily with oxygen. For this reason, silicon chains and hydrogen compounds burst into flame in air, as they break apart to form crystals of silicate. So even if life with silicon were possible, it would not last very long!

Doping silicon

One of the reasons silicon is important to the electronics industry is that it is relatively simple to change how well this element conducts electricity, using a process called doping. Doping can be done in two ways, known as negative (n-type) doping and positive (p-type) doping.

Negative doping

In negative doping, a few atoms—maybe only 10 for every million silicon atoms—are added to the silicon. These can be any atoms that have five electrons in their outer shell. Just as with a silicon atom, four of the new atom's electrons are shared in bonds with neighboring silicon atoms, but now there is a spare fifth electron. This electron is free to roam through the conduction band, carrying electricity.

Positive doping

In positive doping, an atom with only three electrons, such as boron, is added. This atom has a "positive hole," because of its missing electron. It fills this hole by taking an electron from a neighboring silicon atom. This leaves a "positive hole" in that silicon atom, which then takes an electron from another atom, and so on. As a result, a flow of positive holes moves through the silicon, carrying electricity.

DOPING SILICON

Pure silicon

Silicon

Electron

Pure silicon conducts electricity very poorly because each atom shares the four electrons in its outer shell with the four silicon atoms around it.

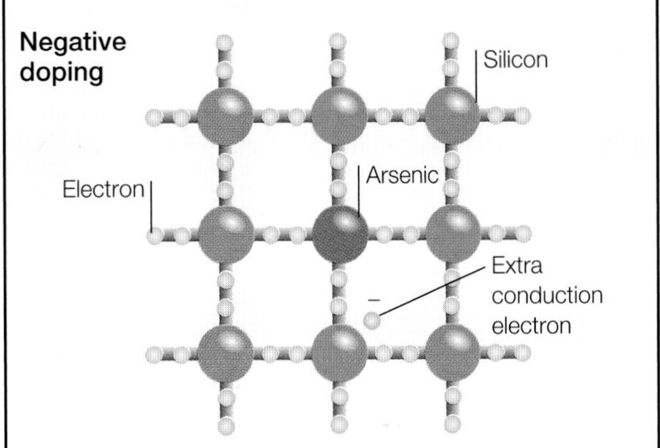

In negative doping, phosphorus atoms are usually used, but arsenic and antimony work just as well. The extra conduction electron carries electricity.

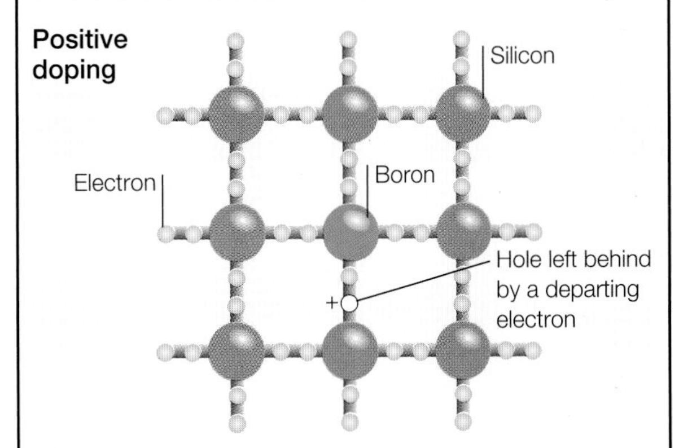

In positive doping, a kind of "musical chairs" takes place, with electrons jumping into the holes created by the departure of their neighbors. Suitable atoms are boron, indium, and gallium.

Silicon chips

The use of silicon in the electronics industry has revolutionized computers. They are so important to us now that it is difficult to imagine life without them.

When the first computers were built, they took up entire rooms and struggled to do calculations that you can do today using a pocket calculator. Silicon has made it possible for scientists to make modern computers that are both small and very powerful.

Silicon as a transistor

The most important device inside a modern computer is a tiny device called a transistor. Applying a small amount of electricity to the transistor causes large changes in the flow of electricity across

DID YOU KNOW?

GIANT CRYSTALS

The wafers that silicon chips are made from are cut from a single giant crystal of silicon. To make these giant crystals, a small "seed" crystal is dipped into a vat of pure liquid silicon. The vat is slowly spun in one direction, while the crystal is turned in the other. This causes the liquid silicon to crystallize onto the seed, which grows bigger as it is gradually pulled out of the vat. Eventually this single crystal may be more than 4 inches (10 centimeters) in diameter and over 5 feet (1.5 meters) long!

"Silicon chips" are also called integrated circuits, because all the electric devices are connected, or integrated, together on a single piece of silicon.

it—just like a gentle turn of a faucet can cause a big change in the flow of water through the faucet. In the early days of computers, before transistors had been invented, vacuum tubes were used. These were much larger than modern transistors and worked by emitting electrons from their metal surfaces when heated. They were not ideal, because they required a certain amount of time to "warm up" and frequently "burned out" when the heating element inside them broke down.

In 1947, scientists discovered that small pieces of n- and p-type semiconductors, such as silicon or germanium, could be used as tiny transistors. Silicon soon became more popular than germanium for this purpose, because silicon is very abundant and silicon oxide is an excellent insulator. The silicon forms all the parts of the transistor, and the oxide can be used as an insulating barrier to separate the transistors from each other.

Integrated circuits

In modern computers, thousands or millions of transistors are incorporated into a single chip. To make chips, technicians apply a mask to a silicon wafer, so that the mask covers and protects certain parts of the wafer but leaves other parts uncovered. The uncovered parts can then have atoms added to them. For example, if a beam of phosphorus atoms is fired at the wafer surface, they penetrate into the silicon and dope it. By using lots of different masks and different elements, it is possible to build a whole integrated circuit on one wafer.

Other uses of silicon

Humans have been using silicon since they appeared on Earth. The Stone Age got its name because people used stone tools and lived in caves in the rock. As we have seen, silicon makes up almost all the rocks and stones around us.

Bricks and glass

We still need silica for our homes today. The bricks that are used for buildings are made by baking clay in an oven at around 1,742 to 2,192°F (950 to 1,200°C) for 12 hours. The cement that holds the bricks together is made from a mixture of calcium carbonate, silica, aluminum, and iron oxide (found in clay).

Silicon is the most important material in glass. Windows and bottles are made from soda glass ($Na_2SiO_3CaSiO_33SiO_2$). Colored glass can be made by adding small quantities of chromium (green), manganese (violet), or cobalt (blue) atoms to the glass as it is made.

Water glass and silica gel

A silica compound called water glass (Na_2SiO_3) is made by heating sand with concentrated sodium hydroxide (NaOH). Water glass is a solid that looks a little like glass. It is so called because it dissolves

Concrete is a very popular building material. It is made by mixing gravel—made mainly of silicates—with cement and water. The huge concrete pipes in this picture will be used for sewers and storm drains.

GLASS FACTS

Soda glass is not the only type of glass that can be made from silicon. Others include:

- pyrex glass—replacing some of the silica in soda glass with boron trioxide forms pyrex glass. This glass is similar to soda glass, but it melts at much higher temperatures. Pyrex glass is used in laboratories all over the world.

- crystal glass—adding small quantities of lead oxide to soda glass makes crystal glass. The lead oxide changes the way the glass reflects light, making it sparkle much more than ordinary glass does. Crystal glass is used to make chandeliers and other ornamental glassware.

easily in water to produce a clear, syrupy liquid. Water glass can be used as a fireproof coating for paper and fireboard, in the manufacture of soaps and detergents, and as a substitute glue used for making cardboard boxes.

Adding acid to water glass forms a gel-like solid. The gel contains a large amount of water, but if it is heated, the water evaporates away. The solid left over after evaporation is called silica gel. Even tiny pieces of this substance have a huge surface area, which allows them to absorb lots of water. For this reason, silica gel is often used in boxes to keep products dry.

Silicon carbide and silicone

Silicon carbide is the hardest artificial substance known. It is made by heating pure sand and carbon (coke) in a furnace. Silicon carbide is used in sandpapers, and as a coating on saws, making the blades so hard that they can cut through solid rock or even gemstones.

Another form of silicon that is extremely useful is silicone. Silicones are rubbery compounds made from long chains of silicon and oxygen atoms, with carbon atoms joined to the silicon atoms. Silicones were first made because people thought they would be much better than

Silicon is a major constituent of soda glass, but adding lead oxide to it creates a sparkling crystal glass, like that used to make this chandelier.

the natural rubber used as the insulating cover for electrical cables. Today silicones have thousands more uses. They are used in space suits, in the gooey liquid used to mend bike punctures, in waterproof clothes, in paints, in suntan lotions, and even in medicines.

This is a retina implant—a tiny silicon chip designed to allow blind people to see. Fixed at the back of the eye in place of the natural retina, the chip releases electrons when it detects light. The electrons stimulate the optic nerve, which runs to the brain.

DID YOU KNOW?

SILICONES

Silicones can withstand high temperatures and are very unreactive. This means they can be heated to make them sterile (germ-free), and they do not react with chemicals in the human body. It is also easy to make both hard and soft silicones. For these reasons, silicones are used in replacement body parts, such as finger and toe implants. They have also been used in breast reconstructions, although the discovery in the 1990s of possible health risks means that implants filled with saltwater are now preferred.

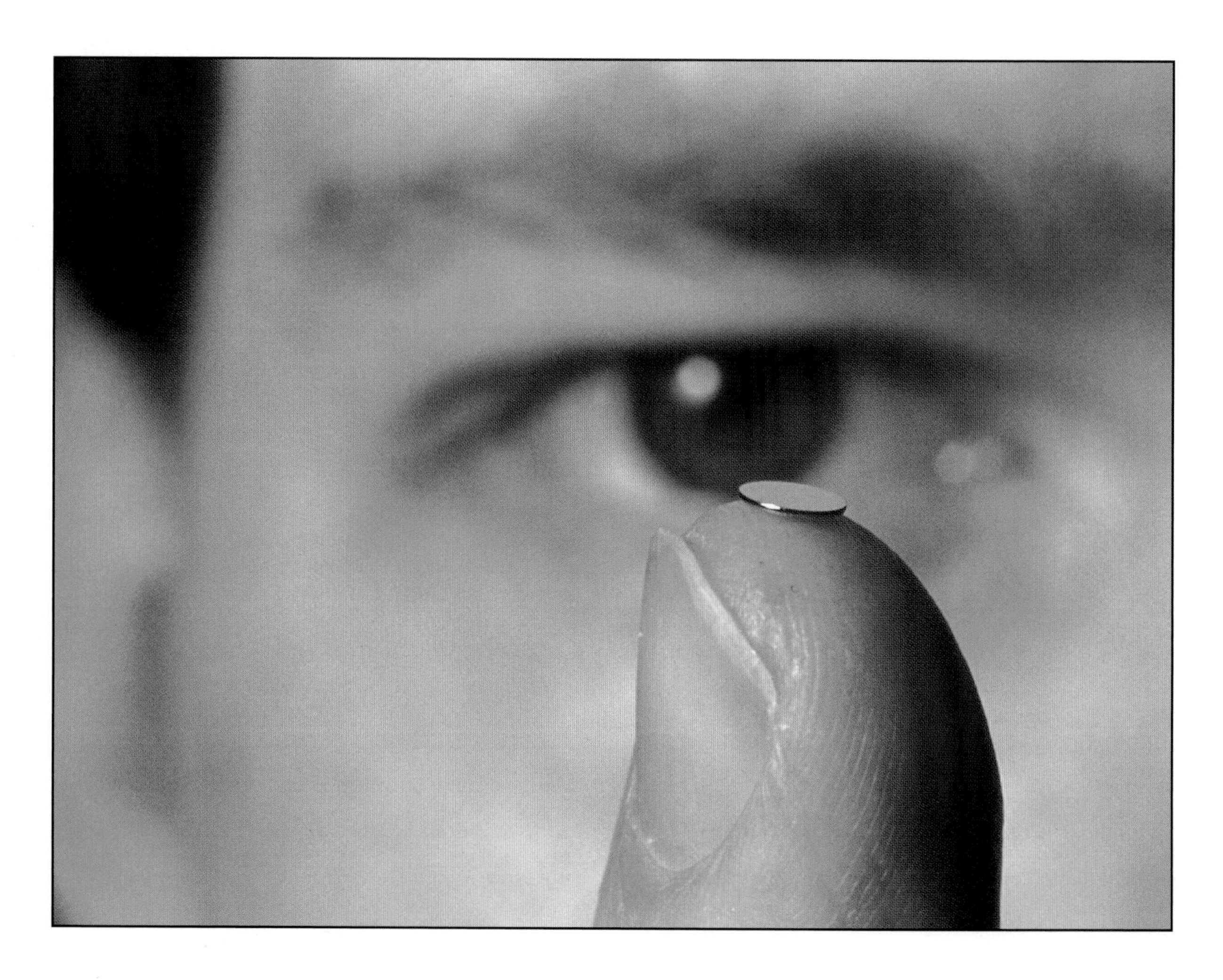

Silicosis

Every year hundreds of people in the United States die from a disease called silicosis. Miners, glassmakers, pottery workers, and other people who breathe in a large amount of silicon dust during the course of their work can develop this disease. Normally it takes many years before silicosis strikes. However, anyone who spends even a few weeks working in a very dusty environment may be at risk.

In the lungs

Silicosis is caused by the tiny particles of silicon dust breathed into the lungs. The silicon attacks lung tissue, causing it to scar and develop growths and preventing the lungs from working properly. Usually, tiny hairs in the throat called cilia push the dust back out. If someone smokes or spends a lot of time working in a dusty environment, however, the cilia stop working and the damage continues.

The first symptoms of silicosis are shortness of breath, coughing, and exhaustion. These symptoms have been recognized for more than 400 years and were known to affect stonecutters in the sixteenth century. If nothing is done to stop it at this stage, the disease may lead to other conditions such as tuberculosis and cancer. Eventually the person dies.

Although silicosis is difficult to treat once it has started, it is easy to prevent in the first place. Wearing a mask, working in a clean, airy environment, and not smoking all reduce the likelihood of developing silicosis.

These miners are wearing face masks to protect themselves from the dusty atmosphere and to prevent the lung damage that characterizes silicosis.

Periodic table

Everything in the Universe is made from combinations of substances called elements. Elements are the building blocks of matter. They are made of tiny atoms, which are much too small to see.

The character of an atom depends on how many even tinier particles called protons there are in its center, or nucleus. An element's atomic number is the same as the number of protons.

Scientists have found more than 110 different elements. About 90 elements occur naturally on Earth. The rest have been made in experiments.

All these elements are set out on a chart called the periodic table. This lists all the elements in order according to their atomic number.

The elements at the left of the table are metals. Those at the right are nonmetals. Between the metals and the nonmetals are the metalloids, which sometimes act like metals and sometimes like nonmetals.

- On the left of the table are the alkali metals. These elements have just one electron in their outer shells.
- Elements get more reactive as you go down a group.
- On the right of the periodic table are the noble gases. These elements have full outer shells.
- The number of electrons orbiting the nucleus increases down each group.
- Elements in the same group have the same number of electrons in their outer shells.
- The transition metals are in the middle of the table, between Groups II and III.

Group I	Group II	Transition metals						
1 H Hydrogen 1								
3 Li Lithium 7	4 Be Beryllium 9							
11 Na Sodium 23	12 Mg Magnesium 24							
19 K Potassium 39	20 Ca Calcium 40	21 Sc Scandium 45	22 Ti Titanium 48	23 V Vanadium 51	24 Cr Chromium 52	25 Mn Manganese 55	26 Fe Iron 56	27 Co Cobalt 59
37 Rb Rubidium 85	38 Sr Strontium 88	39 Y Yttrium 89	40 Zr Zirconium 91	41 Nb Niobium 93	42 Mo Molybdenum 96	43 Tc Technetium (98)	44 Ru Ruthenium 101	45 Rh Rhodium 103
55 Cs Cesium 133	56 Ba Barium 137	71 Lu Lutetium 175	72 Hf Hafnium 179	73 Ta Tantalum 181	74 W Tungsten 184	75 Re Rhenium 186	76 Os Osmium 190	77 Ir Iridium 192
87 Fr Francium 223	88 Ra Radium 226	103 Lr Lawrencium (260)	104 Unq Unnilquadium (261)	105 Unp Unnilpentium (262)	106 Unh Unnilhexium (263)	107 Uns Unnilseptium (?)	108 Uno Unniloctium (?)	109 Une Unillenium (?)

Lanthanide elements	57 La Lanthanum 39	58 Ce Cerium 140	59 Pr Praseodymium 141	60 Nd Neodymium 144	61 Pm Promethium (145)
Actinide elements	89 Ac Actinium 227	90 Th Thorium 232	91 Pa Protactinium 231	92 U Uranium 238	93 Np Neptunium (237)

The horizontal rows are called periods. As you go across a period, the atomic number increases by one from each element to the next. The vertical columns are called groups. Elements get heavier as you go down a group. All the elements in a group have the same number of electrons in their outer shells. This means they react in similar ways.

The transition metals fall between Groups II and III. Their electron shells fill up in an unusual way. The lanthanide elements and the actinide elements are set apart from the main table to make it easier to read. All the lanthanide elements and the actinide elements are quite rare.

Silicon in the table

Silicon has 14 protons in its nucleus, so it has an atomic number of 14. It is a metalloid, or semimetal, and shares Group IV of the periodic table with a nonmetal (carbon), another metalloid (germanium), and two metals (tin and lead). Like these other Group IV elements, silicon has four electrons in its outermost shell.

Metals

Metalloids (semimetals)

Nonmetals

Key	
14	Atomic (proton) number
Si	Symbol
Silicon	Name
28	Atomic mass

			Group III	Group IV	Group V	Group VI	Group VII	Group VIII
								2 He Helium 4
			5 B Boron 11	6 C Carbon 12	7 N Nitrogen 14	8 O Oxygen 16	9 F Fluorine 19	10 Ne Neon 20
			13 Al Aluminum 27	14 Si Silicon 28	15 P Phosphorus 31	16 S Sulfur 32	17 Cl Chlorine 35	18 Ar Argon 40
28 Ni Nickel 59	29 Cu Copper 64	30 Zn Zinc 65	31 Ga Gallium 70	32 Ge Germanium 73	33 As Arsenic 75	34 Se Selenium 79	35 Br Bromine 80	36 Kr Krypton 84
46 Pd Palladium 106	47 Ag Silver 108	48 Cd Cadmium 112	49 In Indium 115	50 Sn Tin 119	51 Sb Antimony 122	52 Te Tellurium 128	53 I Iodine 127	54 Xe Xenon 131
78 Pt Platinum 195	79 Au Gold 197	80 Hg Mercury 201	81 Tl Thallium 204	82 Pb Lead 207	83 Bi Bismuth 209	84 Po Polonium (209)	85 At Astatine (210)	86 Rn Radon (222)

62 Sm Samarium 150	63 Eu Europium 152	64 Gd Gadolinium 157	65 Tb Terbium 159	66 Dy Dysprosium 163	67 Ho Holmium 165	68 Er Erbium 167	69 Tm Thulium 169	70 Yb Ytterbium 173
94 Pu Plutonium (244)	95 Am Americium (243)	96 Cm Curium (247)	97 Bk Berkelium (247)	98 Cf Californium (251)	99 Es Einsteinium (252)	100 Fm Fermium (257)	101 Md Mendelevium (258)	102 No Nobelium (259)

Chemical reactions

Chemical reactions are going on all the time—candles burn, nails rust, food is digested. Some reactions involve just two substances; others many more. But whenever a reaction takes place, at least one substance is changed.

In a chemical reaction, the atoms stay the same. But they join up in different combinations to form new molecules.

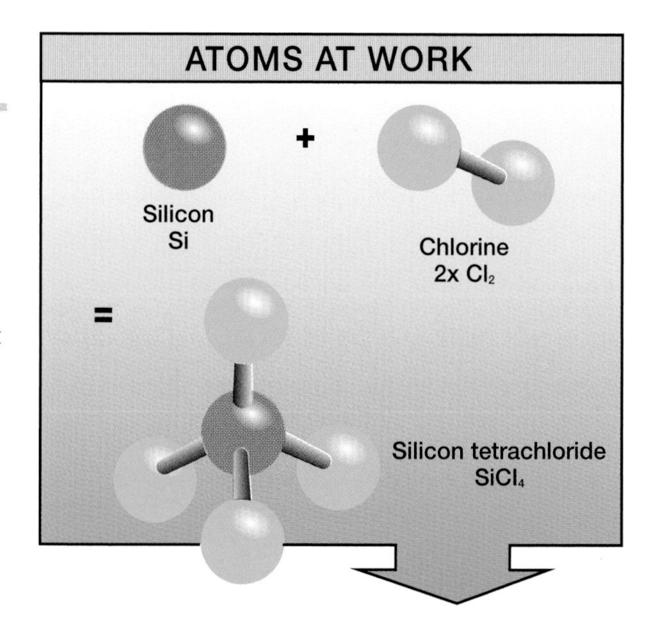

The reaction that takes place when silicon reacts with chlorine can be written like this:

$$Si + 2Cl_2 \rightarrow SiCl_4$$

This shows us that one atom of silicon reacts with two molecules of chlorine to form one molecule of silicon tetrachloride.

Writing an equation

Chemical reactions can be described by writing down the atoms and molecules before and the atoms and molecules after. Since the atoms stay the same, the number of atoms before will be the same as the number of atoms after. Chemists write the reaction as an equation. The equation shows what happens in the chemical reaction.

When the numbers of each atom on both sides of the equation are equal, the equation is balanced. If the numbers are not equal, something is wrong. So the chemist adjusts the number of atoms involved until the equation balances.

Armies have developed silicon tetrachloride for use as a "smoke screen" during war, to make it difficult for their enemies to see what they are doing.

Glossary

atom: The smallest part of an element that has all the properties of that element.
atomic mass: The number of protons and neutrons in an atom.
atomic number: The number of protons in an atom.
bond: The attraction between two atoms that holds them together.
compound: A substance that is made of atoms of more than one element. Silica (silicon oxide) is a compound of silicon and oxygen.
conductor: A substance through which heat or electricity flow easily.
crystal: A solid substance in which the atoms are arranged in a regular three-dimensional pattern.
density: The mass of a substance in a given volume.
doping: A process in which the electrical conductivity of pure silicon is increased by adding atoms of another element to it.
electron: A tiny particle with a negative charge. Electrons are found inside atoms, where they move around the nucleus in orbits called electron shells.
element: A substance that is made from only one type of atom.
evaporation: A process in which liquid turns to vapor. Evaporation takes place at a temperature below the boiling point.
insulator: A substance that is a poor conductor of heat or electricity.
isotopes: Atoms of an element with the same number of protons and electrons but different numbers of neutrons.
metal: An element on the left-hand side of the periodic table.
Mohs scale: A scale of hardness used to classify minerals.
molecule: A particle that contains atoms held together by chemical bonds.
neutron: A tiny particle with no electrical charge. It is found in the nucleus of almost every atom.
nonmetal: An element on the right-hand side of the periodic table.
nucleus: The center of an atom. It contains protons and neutrons.
periodic table: A chart of all the chemical elements laid out in order of their atomic number.
proton: A tiny particle with a positive charge. Protons are found inside the nucleus of an atom.
refining: An industrial process that frees elements, such as metals, from impurities or unwanted material.
semiconductor: A substance through which heat or electricity flow more easily than through an insulator but less easily than through a conductor.
transistor: A small device used in electronics that contains a semiconductor and is used as a switch or amplifier.

Index